SCIENTIFIC AMERICAN | EDUCATIONAL PUBLISHING

SCIENTIFIC AMERICAN INVESTIGATES CAREERS IN SCIENCE

GEOLOGIST

BY ROSIE BANKS

Published in 2025 by The Rosen Publishing Group
in association with Scientific American Educational Publishing
2544 Clinton Street, Buffalo NY 14224

Copyright © 2025 Rosen Publishing Group

Library of Congress Cataloging-in-Publication Data
Names: Banks, Rosie, 1978- author.
Title: Geologist / Rosie Banks.
Description: New York : Scientific American Educational Publishing, an imprint of Rosen Publishing, [2025] | Series: Scientific American investigates careers in science | Includes index.
Identifiers: LCCN 2024036202 (print) | LCCN 2024036203 (ebook) | ISBN 9781725352353 (library binding) | ISBN 9781725352346 (paperback) | ISBN 9781725352360 (ebook)
Subjects: LCSH: Geology–Vocational guidance–Juvenile literature. | Geologists–Juvenile literature. | Geology–Juvenile literature.
Classification: LCC QE34 .B36 2025 (print) | LCC QE34 (ebook) | DDC 551.023–dc23/eng20240814
LC record available at https://lccn.loc.gov/2024036202
LC ebook record available at https://lccn.loc.gov/2024036203
Designer: Leslie Taylor
Editor: Therese Shea

Portions of this work were originally authored by Ryan Nagelhout and published as *Be an Geologist*. All new material in this edition is authored by Rosie Banks.

Photo credits: Cover (main) Yes058 Montree Nanta/Shutterstock.com; series art (background) jijomathaidesigners/Shutterstock.com; p.4 yoganugraha/Shutterstock.com; p. 5 Pornpimon Ainkaew/Shutterstock.com; p. 7 (main) Krystalp20/Shutterstock.com, (inset) National Galleries Scotland/File:Sir Henry Raeburn - James Hutton, 1726 - 1797. Geologist - PG 2686 - National Galleries of Scotland.jpg_commons.wikimedia.org; p. 8 dave souza/File:Siccar Point red capstone closeup.jpg_commons.wikimedia.org; p. 9 Daboost/Shutterstock.com; p. 10 Christopher Edwin Nuzzaco/Shutterstock.com; p. 11 (main) Wead/Shutterstock.com, (granite) michal812/Shutterstock.com, (pumice) Vasylchenko/Shutterstock.com; p. 13 Wisanu Boonrawd/Shutterstock.com; p. 15 Kavic.C/Shutterstock.com; p. 17 M.Bitton/File:Tectonic plates (2022).svg_commons.wikimedia.org; p. 19 Radu Razvan/Shutterstock.com; p. 20 Peltoms~commonswiki/File:Snowpitglacier.jpg_commons.wikimedia.org; p. 21 SolStock/iStockphoto.com; p. 23 salajean/Shutterstock.com; p. 25 chinaface/iStockphoto.com; p. 26 NASA/File:Harrison Schmitt cropped.jpg_commons.wikimedia.org; p. 27 NASA/JPL-Caltech/File:Perseverance-Selfie-at-Rochette-Horizontal-V2.gif_commons.wikimedia.org; p. 29 (background) olko1975/Shutterstock.com, (3 illustrations) AI-generated images/Shutterstock.com.

Some of the images in this book illustrate individuals who are models. The depictions do not imply actual situations or events.

All rights reserved. No part of this book may be reproduced in any form without permission in writing from the publisher, except by a reviewer.

Printed in the United States of America

CPSIA compliance information: Batch #CWSA25. For Further Information contact Rosen Publishing at 1-800-237-9932.

Find us on

CONTENTS

Words in the glossary appear in **bold** type the first time they are used in the text.

AN EXCITING CAREER

Geology is the study of Earth. Would you want to be a geologist? You could study Earth as it is now or try to find out more about Earth in the past. You'll discover that Earth has changed—a lot! And you can also discover why.

Geologists are key to our understanding of our planet. We need to know as much as we can about Earth. This will help us care for it now. It also helps us predict, or guess, what may happen in the future. If that sounds like an exciting career to you, that's just the beginning of what a geologist does!

FUN FACT

THE WORD "GEOLOGY" COMES FROM THE LATIN WORD *GEOLOGIA*. IT MEANS "THE STUDY OF EARTH."

Geologists know that rocks are the key to many of Earth's mysteries!

THE FATHER OF GEOLOGY

Many people have had different ideas about how Earth's geological processes worked. Scottish farmer James Hutton (1726–1797) is called the father of geology for his **theory** that Earth is always re-forming its surface. Hutton went to college to study chemistry and medicine, but he became a farmer. He noticed how weather changed his farmland year after year.

Many people during Hutton's time believed Earth was only about 6,000 years old. Hutton suggested the planet was older than that. He thought the processes shaping the land had been working for much longer than a few thousand years.

FUN FACT

GEOLOGISTS TODAY THINK EARTH IS 4.5 BILLION YEARS OLD!

James Hutton

While powerful events do change Earth's landforms, they aren't the only causes of change at work.

Slow and Steady

Hutton's most famous theory is uniformitarianism. That long word is the idea that Earth is being shaped by the same gradual processes today as it always has been. These changes happen at the same slow rate as well. Before Hutton's theory, most people believed quick, powerful events such as floods formed and changed Earth. This theory was called catastrophism.

Hutton used his own observations of the world around him as evidence, or proof, of his theory. He observed that rocks formed in layers, with different colors and types lying on top of each other. He noted how slowly these layers take shape over time. He also described how they form and break down. He called it a "great geological cycle."

Hutton thought changes on Earth were caused by heat and pressure under the ground. Heat and pressure cause chemical changes in matter. Heat and pressure also cause movement of pieces of Earth's outer layer, or crust.

Siccar Point

Volcanic eruptions and hot springs are evidence of the great heat within Earth.

Pointing to History

Hutton used rocks at Siccar Point in Scotland to support his theories. Siccar Point is a landform jutting into the North Sea that features two different types of rock layers slanting in different directions. Its makeup suggested to Hutton that the layers formed at different time periods in Earth's history.

THE ROCK CYCLE: IGNEOUS

James Hutton also developed the idea of the rock cycle. There are three main types of rocks: igneous, sedimentary, and metamorphic. Each can change into the other two forms, depending on conditions around them.

Igneous rocks are formed from magma, which is hot, liquid rock inside Earth. Magma flows up through cracks in the ground. It may get trapped in hollow pockets where it slowly cools to become igneous rock with large crystals. If magma reaches the surface, it's called lava. Lava quickly cools to form a more lightweight igneous rock with smaller or no crystals.

FUN FACT

THE IGNEOUS ROCK PUMICE IS SO LIGHT THAT IT CAN FLOAT!

pumice

Granite is a kind of igneous rock that forms underground. Pumice is a kind of igneous rock that forms aboveground.

granite

pumice

Wanted: Volcanologists

Volcanologists are geologists who study volcanoes and the processes that take place under Earth to make them erupt. They try to figure out when volcanoes may erupt, which can save lives. Some volcanologists study underwater volcanoes too! Since volcanic rock can be full of valuable **minerals**, many businesses are interested in hiring these geologists.

THE ROCK CYCLE: SEDIMENTARY

The other two kinds of rocks in the rock cycle form much more slowly. Sedimentary rocks are made of bits of rock, sand, shells, and other matter. Sometimes these pieces are created by weathering, a process in which water, ice, snow, and wind break down rocks into smaller parts. Wind, water, or ice may carry away the bits and deposit, or place, them in another area. Over time, they collect, **compact**, and harden, or cement, into rock.

Sedimentary rock also forms from bits of shells that collect on the ocean floor. In addition, sedimentary rock can form from minerals left behind after water **evaporates**.

FUN FACT

FOSSILS ARE FOUND ONLY IN SEDIMENTARY ROCKS!

Weathering and **erosion** formed the Grand Canyon over millions of years.

THE ROCK CYCLE: METAMORPHIC

Metamorphic rocks form from sedimentary or igneous rocks that experience great heat or pressure inside Earth. This causes the rock to undergo a process of change. Heat and pressure can alter a rock's **texture**, appearance, and even chemical makeup. Certain liquids and gases can also add to these changes.

Metamorphic rocks often have waves of minerals in different colors. As the rocks change, the minerals within them separate to form "ribbons." The rocks may have shiny crystals in them too. Limestone, a sedimentary rock, can become the metamorphic rock marble.

Metamorphic rocks come to the surface through weathering, erosion, or other natural processes. Then, the rock cycle continues.

FUN FACT

METAMORPHIC ROCKS CAN ALSO COME TO THE SURFACE THROUGH UPLIFT. THIS IS THE PROCESS BY WHICH EARTH'S CRUST GETS PUSHED UP. UPLIFT EXPLAINS HOW MOUNTAINS GROW TALLER!

As this chart shows, the rock cycle moves in all directions.

ROCK CYCLE

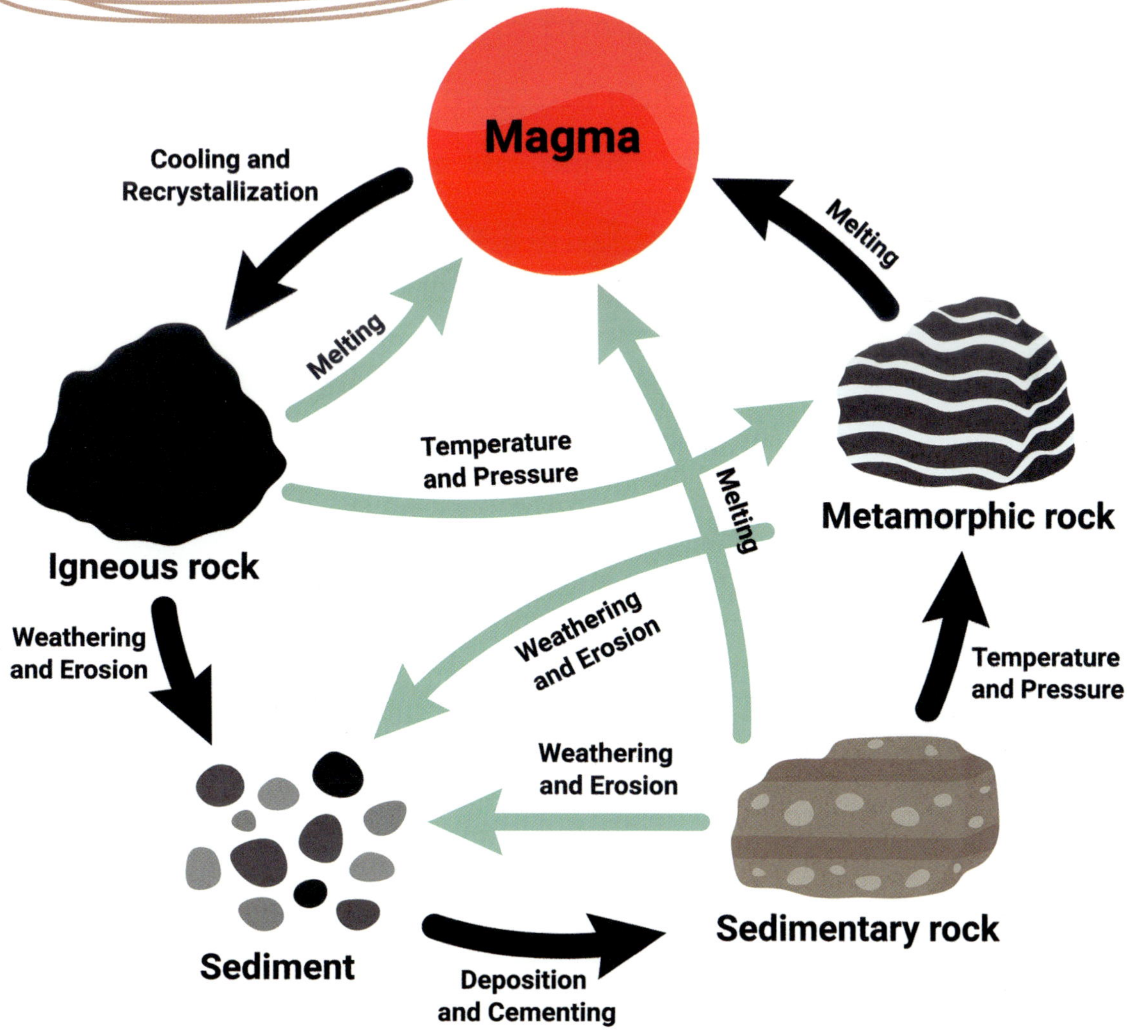

IT'S A PUZZLE!

In 1912, a scientist named Alfred Wegener suggested a theory that Earth's continents had drifted, or moved slowly, apart over time. He didn't receive much support, however. After all, how could continents move?

Later, geologists realized Earth's outer layer works like a puzzle. It's made up of broken pieces, or plates, of different sizes that slowly move. There are dozens of plates, and the major continents rest on the biggest ones.

These plates grind past each other and cause **earthquakes**. Some push together to create mountains. They move continents closer together or farther apart. The plates move about 2 to 4 inches (5 to 10 cm) a year.

The theory of Earth's outer layer consisting of many moving pieces is called plate tectonics. The crust under the continents is usually about 25 miles (40 km) thick. Oceanic crust is about 4 miles (6 km) thick.

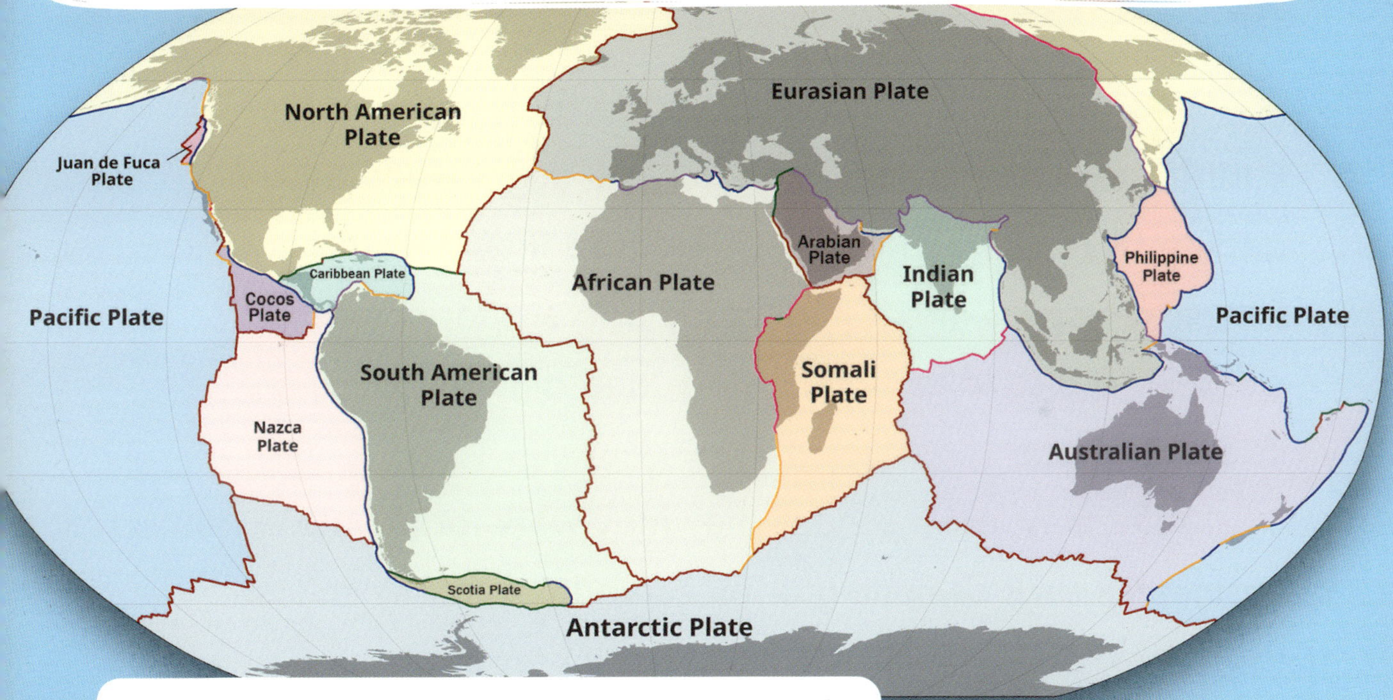

Interested in Earthquakes?

Some geologists study earthquakes. They're called seismologists. Paleoseismology is the study of prehistoric earthquakes, or earthquakes before written records. Paleoseismologists study the timing, location, and size of these earthquakes. They look for prehistoric faults, or breaks in Earth's crust. These may hold prehistoric rock and soil that tell us what Earth was like thousands of years ago.

SCHOOL, THEN WORK

Are you interested in what you've read so far about Earth? Do you like being outdoors? You might want to be a geologist! You'll need a college **degree** first, though. Future geologists major in geology or a similar course of study. They also take math, history, and different kinds of science classes.

After they leave school, they may work for private businesses such as mining or oil companies. Many choose to work for governments.

Some positions may require geologists to have a higher degree, which means more years of school. Many geologists become **researchers**. They specialize, or focus, their studies and try to figure out a geological problem or mystery.

Many geologists do fieldwork, or study at an outdoor site, at least part of the time. The rest of their time is spent in science labs, classrooms, or offices preparing reports and using computers.

EARTH IS A CLASSROOM

Don't let years of studying discourage you from a geology career. Geology professors try to get their students outside when they can. They take their classes into the field to study different kinds of land formations, show proof of Earth's processes, and observe changing surroundings to learn about an area's geologic history. Sometimes, this means traveling to amazing locations around the world!

Some schools ask geology teachers and students to research and publish their findings. Teaching is a career path for geologists who want to spread their love for Earth science to others and keep learning about the planet themselves.

The need for geologists and similar Earth scientists is growing, which is good news for geologists seeking a job.

Icy Evidence

Another form of geology doesn't focus on rocks. Glacial geologists, or glaciologists, study ice, including snow, sea ice, glaciers, and ice sheets. They examine how snow and ice affect Earth today as well as how they affected it in the past. Glaciologists may work in places such as Antarctica, Greenland, and the Arctic.

MINING THE EARTH

Mines are the source of many things that people need or want, and geologists can find underground treasures. Geologists who work for mining companies locate valuable minerals, called ore, in the ground. They also figure out the best way to remove them. Ore may be valuable itself or valuable for what it contains, such as the elements copper, silver, or gold.

Geologists help plan methods for workers to remove ore safely. In addition, they try to help the company lessen a mine's harm to the **environment**. Mines can have a big effect on plants and animals in the area.

It can be costly to dig a mine, so a geologist figures out the best areas to find ore of the highest value.

Hide-and-Seek Elements

Much of the **technology** we use today, such as TVs, smartphones, and computers, contain what are called rare earth elements. They aren't rare, but they aren't often found in large amounts. Without a geologist, mining companies wouldn't know where or how deep to dig, or when to stop looking for these elements.

ENERGY EMPLOYEES

Natural gas, coal, and oil are energy sources. They form over millions of years from plant and animal remains buried deep within Earth. That's why they're called fossil fuels. The kind of fuel produced is decided by where the matter is forming underground, the temperature there, and the matter surrounding it.

Geologists find new sources of oil and gas by collecting samples of sedimentary rock to analyze, or study closely. They may drill into the earth to get these samples. This may mean drilling into the ocean floor! Geologists not only find the source but also try to figure out the amount that's there.

Geologists who search for oil and gas deposits may be called petroleum geologists. Oil and gas are often found together.

Renewable Energy

Fossil fuels are called nonrenewable energy. They have a limited supply and cannot be replaced. Some geologists specialize in renewable energy sources. For example, they locate geothermal energy sources where heat is escaping from within Earth. Geothermal energy can be used for heat and for producing electricity.

NOT JUST EARTH SCIENCE

Geologists help us understand Earth—and they can help us understand other planets too! Geologists who work for NASA (National Aeronautics and Space Administration) study planets as well as moons, **asteroids**, and other space objects. After all, these are made of the same matter as our own planet.

Scientists are finding out a lot about Mars with the help of robotic vehicles called rovers. Rovers move around the planet's surface, take photos, and collect rock samples. They can analyze the rocks and send findings back to Earth.

FUN FACT

ASTRONAUT HARRISON SCHMITT WAS THE ONLY GEOLOGIST TO VISIT THE MOON. HE WAS ON THE APOLLO 17 SPACE MISSION OF 1972.

The Perseverance rover on Mars collected igneous rocks that showed evidence of water on the planet.

LET'S GET STARTED!

Geologists are never far from their work. Everywhere they look is something to study. The hardest part of becoming a geologist might be figuring out what to study. This career can take you around the world or even off planet. It depends on your interests.

You can explore a geology career right now, no matter your age. Museums are a great way to meet and learn from real-life geologists. You can also join a club for people interested in the geology of your area. You might find evidence of your hometown's history at the beach or a local park. Anyone might make a geological discovery!

More Kinds of Geologists

Name	What Do They Do?
engineering geologist	studies the ground under construction sites to determine the safety of building
geochemist	studies the chemicals found within rocks and minerals
environmental geologist	studies how people and the environment affect each other
marine geologist	studies minerals and rocks in and under the ocean
geomorphologist	studies how natural events and processes change lands
geohydrologist	studies water under the ground
geophysicist	studies electrical, magnetic, and physical properties of Earth

What features of our amazing planet interest you? There's a kind of geologist who studies it!

GLOSSARY

asteroid: An object in space smaller than a planet made of mostly rock and metal that orbits the sun.

compact: To force closer together.

degree: An official paper given to someone who has completed certain classes at a college or university.

earthquake: A shaking of the ground caused by the movement of Earth's crust.

environment: All the conditions that surround something and affect it.

erosion: The act of wearing away by water, wind, or ice.

evaporate: To change from a liquid to a gas.

mineral: Matter in the ground that forms rocks.

researcher: One who studies to find out something new.

technology: A machine, piece of equipment, or method created by science, engineering, and other industries.

texture: The structure, form, and feel of something.

theory: A general idea, based on facts, that explains something.

volcanic eruption: The bursting forth of hot, liquid rock from within the earth.

FOR MORE INFORMATION

Books

Bradley, Doug. *20 Things You Didn't Know About Geology*. Buffalo, NY: PowerKids Press, 2023.

Dodd, Emily. *Geology*. New York, NY: Kingfisher, 2022.

Moening, Kate. *Geologist*. Minneapolis, MN: Bellwether Media, 2023.

Websites

Geology for Kids
www.sciencekids.co.nz/geology.html
Science for Kids has videos and games to highlight important geological concepts.

Geology 101
kids.nationalgeographic.com/science/article/geology-101
National Geographic Kids offers more information on geology.

Geology STEM Activities for Kids
www.sciencebuddies.org/stem-activities/subjects/geology
Check out these fun geology experiments!

Publisher's note to educators and parents: Our editors have carefully reviewed these websites to ensure that they are suitable for students. Many websites change frequently, however, and we cannot guarantee that a site's future contents will continue to meet our high standards of quality and educational value. Be advised that students should be closely supervised whenever they access the internet.

INDEX